I0493862

Electromagnetic Compatibility Between Marine Automatic Identification and Public Correspondence Systems in the Maritime Mobile VHF Band

technical report

U.S. DEPARTMENT OF COMMERCE • National Telecommunications and Information Administration

NTIA REPORT 00-376

Electromagnetic Compatibility Between Marine Automatic Identification and Public Correspondence Systems in the Maritime Mobile VHF Band

Robert L. Sole
Brent Bedford
Gary Patrick

U.S. DEPARTMENT OF COMMERCE
William Daley, Secretary

Greg Rohde, Assistant Secretary
for Communications and Information

April 2000

Executive Summary

The Coast Guard funded the National Telecommunications and Information Administration (NTIA) to perform electromagnetic compatibility (EMC) tests between an ITU-R M. 825-3 (*Characteristics Of a Transponder System Using Digital Selective Calling Techniques for Use with Vessel Traffic Services and Ship-to Ship Identification*) based Automatic Identification System (AIS) operating on 12.5 kHz channels and Public Correspondence (PC) Systems operating on 25 kHz channels. The tests were performed between January 17-28, 2000 in and around an AIS base station communications tower located at Point Ala Hache, La. by NTIA, SETA Corporation, and Coast Guard personnel.

AIS is a shipborne transponder-based navigation safety system that serves as the foundation for the Vessel Traffic Service (VTS) being established in New Orleans and elsewhere by the Coast Guard under the Ports And Waterway Safety System (PAWSS) project. AIS is based on technical standards established by the International Telecommunication Union (ITU). The current implementation in New Orleans is based on ITU-R Recommendation M.825-3. This will be updated in the future to the standard for Universal Shipborne AIS, ITU-R M. 1371 (*Technical Characteristics for a Universal Shipborne Identification System Using Time Division Multiple Access In the VHF Maritime Mobile Band*). Transponders that are fully compliant with ITU-R M. 1371 are not currently available. AIS facilitates the efficient exchange of data between ships and between shore stations and ships. AIS responds to the mariners' need for timely, relevant and accurate information (including data on ships position, speed, etc) delivered in an unobtrusive manner. AIS requires dedicated frequencies in order to operate safely and reliably.

AIS requires duplex channels for ship-to-shore and shore-to-ship digital data transmissions and simplex channels for ship-to-ship operations. Frequencies selected for AIS must come from Appendix 18 of the International Radio Regulations (between 156.025-157.425 MHz and 160.625-162.025 MHz).

Of the 35 duplex channels listed in Appendix 18 of the ITU Radio Regulations, only nine remain for consideration for potential AIS operations in the United States due to past U.S. regulatory decisions. These nine 25 kHz duplex channels are currently utilized within the VHF marine Public Correspondence (PC) Band, and designated as channels 24, 84, 25, 85, 26, 86, 27, 87, and 28. To obtain additional AIS frequencies, the U.S. must utilize techniques outlined in ITU-R M. 1084-3 (*Interim Solutions for Improved Efficiency In The Use of the Band 156-174 MHz by Stations In the Maritime Mobile Service*) regarding the use of 12.5 kHz interstitial channels that are interleaved between existing 25 kHz wideband channels.

AIS and Public Correspondence systems both use duplex channels in the maritime mobile VHF band for communications between a mobile unit and a base station, which results in five interference scenarios occurring between the two systems. The four interference scenarios tested were: 1) an AIS base station causing interference to a PC mobile radio receiver; 2) a PC base station causing interference in a AIS transponder receiver; 3) an AIS transponder causing interference in a PC base station receiver; and 4) a PC mobile radio causing interference to an AIS base station receiver. At a minimum, 12.5 kHz of frequency separation was used in all tests scenarios between the interfering transmitter and victim receiver. In addition, 25 kHz, 37.5 kHz, 50 kHz and 62.5 kHz frequency separations were also tested. Tests at 25 kHz and 50 kHz frequency separations are not be applicable since they would require that wideband Public Correspondence to operate on interstitial channels. These tests were performed to obtain additional results to establish data trends. A fifth test scenario would be required to determine compatibility between ship-to-ship AIS and PC operations, and this was not tested due to the unavailability of suitable equipment. This fifth scenario represents a co-site concern with both AIS and PC systems operating on the same ship. The inability to test the fifth scenario does not alter the conclusions or recommendations given in this report.

Analyses of the results of the four test scenarios offer general guidelines for determining compatibility between AIS operations on interstitial 12.5 kHz channels and Public Correspondence operations on wideband 25 kHz channels.

Analyses of the test results show that, since AIS and PC systems both offer service to mariners on ships and would employ base stations with transmission towers located in the same geographic environment, operating these systems is not practical with 12.5 kHz of frequency separation (i.e., geographical separation distances on the order of 15 to 25 miles are required). Operating the systems in the same geographic environment with a frequency separation of 25 kHz may be possible if the PC system (base stations and mobile radios) were designed for 12.5 kHz channel operations. This would require testing of suitable equipment to verify this specific case. There are no current plans to modify the PC system for 12.5 kHz operation. The PC and AIS systems should be able to operate within the same geographic environment provided that a minimum of 37.5 kHz of frequency separation is provided between the two systems.

NTIA recommends that the Coast Guard consider: 1) Developing an AIS frequency coordination plan for the lower Mississippi River for the PC and AIS systems that will ensure mutually compatible and satisfactory operations. 2) Performing additional EMC testing between ship-to-ship AIS and PC operations. 3) Performing EMC tests between PC systems and ITU-R. M 1371 compliant AIS equipment when such equipment becomes available, 4) Pursuing necessary regulatory changes to improve AIS and PC operations in the same geographical area (e.g., including a 12.5 kHz channelization plan for both AIS and PC operations and developing appropriate receiver standards).

CONTENTS

SECTION 1
INTRODUCTION

1.1 Background

The Coast Guard plans to operate an Automatic Identification System (AIS) Digital Selective Calling (DSC) based transponder system as part of the Ports and Waterways Safety System (PAWSS) in the lower Mississippi River. The PAWSS utilizes a combination of voice and AIS working channels in the VHF maritime mobile band to provide communications in a defined Vessel Traffic Service Area (VTSA). The current AIS operates utilizing protocols established by the International Telecommunication Union (ITU) in Recommendation ITU-R M.825-3 (*Characteristics Of a Transponder System Using Digital Selective Calling Techniques for Use with Vessel Traffic Services and Ship-to Ship Identification*). In the future the system will transition to ITU-R Recommendation M.1371 (*Technical Characteristics for a Universal Shipborne Identification System Using Time Division Multiple Access In the VHF Maritime Mobile Band*) protocol when equipment utilizing this new standard is available. This recommendation is also known as the Universal AIS standard, which provides safety and efficiency enhancements over the previously approved ITU-R M.825.3.

Frequencies selected for AIS operations must come from Appendix 18 of the International Radio Regulations (between 156.025-157.425 MHz and 160.625-162.025 MHz). The AIS system requires two or more full duplex channels for ship-to-shore and shore-to-ship digital data transmissions. AIS also utilizes simplex channels(s) for ship-to-ship communications. A more detailed explanation of AIS operations is contained in Appendix A. Of the 35 duplex channels listed in Appendix 18 of the ITU Radio Regulations, only nine remain available for consideration for potential AIS operations in the United States due to past U.S. regulatory decisions. These nine 25 KHz duplex channels are currently utilized for the maritime Public Correspondence (PC) Service, and designated as channels 24, 84, 25, 85, 26, 86, 27, 87, and 28. To obtain additional spectrum for AIS, the U.S. is considering utilizing the techniques outlined in ITU-R M. 1084-3 (*Interim Solutions for Improved Efficiency in the Use of the Band 156 – 174 MHz by Stations In the Maritime Mobile Service*) to interleave 12.5 KHz channels between existing 25 kHz wideband channels. The necessary channel numbering is provided in ITU-R M. 1084-3.

The current AIS duplex working channels, self-designated as 90 and 94, are not listed as duplex pairs in Appendix 18 of the ITU Radio Regulations. The ship-to-shore side of these working channels is shared with other authorized users. Interference generated by these users reduces the data throughput on the ship-to-shore link of the AIS. Therefore, to reduce interference to the AIS and operate the system on channels listed in Appendix 18, the Coast Guard is investigating selecting interstitial channels listed in Appendix 18. The frequencies of the channels are shown in Table 1-1.

Table 1-1
Duplex Channel Frequencies

Channel Designation	Transponder Transmit Frequency (MHz) A Side	Base Station Transmit Frequency (MHz) B Side
283	157.1875	161.7875
24	157.2000	161.8000
224	157.2125	161.8125
84	157.2250	161.8250
284	157.2375	161.8375
25	157.2500	161.8500
225	157.2625	161.8625
85	157.2750	161.8750
285	157.2875	161.8875
26	157.3000	161.9000
226	157.3125	161.9125
86	157.3250	161.9250
286	157.3375	161.9375
27	157.3500	161.9500
227	157.3625	161.9625
87	157.3750	161.9750
287	157.3875	161.9875
28	157.4000	162.0000
228	157.4125	162.0125

AIS channel selections and operations on them must consider existing local PC license holders to preclude mutual interference from occurring between the two systems. Channels selected for these tests were 87, 287, 27, and 227. The choice of these frequencies was arbitrary. Any channel selected from Table 1-1 for the tests would be sufficient, since all adjacent 12.5 kHz and 25 kHz channels have the same characteristics with respect to each other. Channels identified in Table 1-1 by 3 digits are known as narrowband (or interstitial) 12.5 kHz channels and channels that are identified by 2 digits are known as wideband 25 kHz channels. This labeling is in accordance with ITU-R M.1084-3. For these tests, AIS operations were on 12.5 kHz channels and Public Correspondence operations on 25 kHz channels.

The Coast Guard funded NTIA to perform electromagnetic compatibility tests on the AIS and PC systems and to determine separation distances necessary to preclude mutual interference from occurring between the two systems. The tests were performed January 17-28, 2000 in and around the PAWSS tower two site located at Point Ala Hache, Louisiana by NTIA, SETA, and Coast Guard personnel. The tower is located at N29-34-50/W89-49-40. The AIS antennas were mounted 111 meters above ground level. The results of these tests are given in the following sections.

A spectrum snapshot of the emitters that were active during the testing within the VHF maritime mobile band was taken with the a spectrum analyzer connected to the tower two receive antenna. The plots are shown in Appendix B. It should be noted that the snapshot is not representative of all emitters that could be active in the area.

1.2 Test Objectives

The objectives were to investigate the four interference scenarios outlined below and determine separation distances between base and mobile units of the PC and AIS systems to minimize mutual interference from occurring between the two systems. Both closed loop and radiated tests were performed.

Scenario 1: AIS base station transmitter causing interference in a mobile Public Correspondence users radio receiver.

Scenario 2: Public Correspondence base station transmitter causing interference in an AIS transponder receiver.

Scenario 3: AIS transponder transmitter causing interference in a Public Correspondence base station receiver.

Scenario 4: Public Correspondence mobile transmitter causing interference in an AIS base station receiver.

SECTION 2
TEST PROCEDURES

2. Performance Objectives

The performance of the PC base station and mobile radios was based on SINAD measurements that were performed with a communications test set and the aural quality of the received signal which was judged by a listener. A SINAD measurement, in dB, is the ratio of the desired signal to the desired signal added to interference, noise and distortion. A SINAD measurement of 12 dB and above is usually considered adequate for communications. A SINAD degradation to 14 dB from adjacent channel AIS interference was the performance objective for the receivers in the PC base station and mobile radios for these tests. This level is consistent with the specifications of the International Electrotechnical Commission (IEC) document 1097-7 and the Radio Technical Commission for Maritime Services (RTCM) document 87-99[1] for allowable degradation to an analog FM marine radio receiver. For aural measurements, the criteria was that the audible AIS data bursts could be no louder or discernable than the normal background noise in the receiver. This test was subjective and depended on the hearing capabilities of the listener.

The performance of the AIS base station and transponder was based on a received report count. A received report count is the number of AIS reports that a AIS receiver could demodulate and decode in a fixed amount of time, versus the expected number of received reports. For these tests the minimum rate was 90 percent. For example, in a two minute period with a 6 second reporting rate the expected received report count is 20 reports. The allowable degradation would be 2 reports. The 90 percent rate was chosen to ensure a high data throughput and still allow some degradation to the receiver performance from adjacent channel interference. Reporting rates from 1 to 6 seconds were used during the tests.

2.1 Test Procedures

The procedures for the closed loop and radiated tests for each of the four scenarios are given in the following sections of this document.

2.2 Scenario 1
2.2.1 Scenario 1 Closed Loop Test Procedures

Two mobile Public Correspondence (PC) radio receivers were tested for susceptibility to interference from AIS base station transmitter emissions using the following procedures. Commercial and recreational grade VHF radios were tested. Radio A is a commercial grade radio that is Global Maritime Distress and Safety System (GMDSS) and RTCM SC-117 certified. Radio B is representative of the type used by recreational boaters. A diagram of the test set-up is shown below in Figure 2-1.

---

[1]"RTCM Recommended Standards For Installed Maritime VHF Radiotelephone Equipment Operating In High Level Electromagnetic Environments", Radio Technical Commission for Maritime Services (Alexandria Va., 1999).

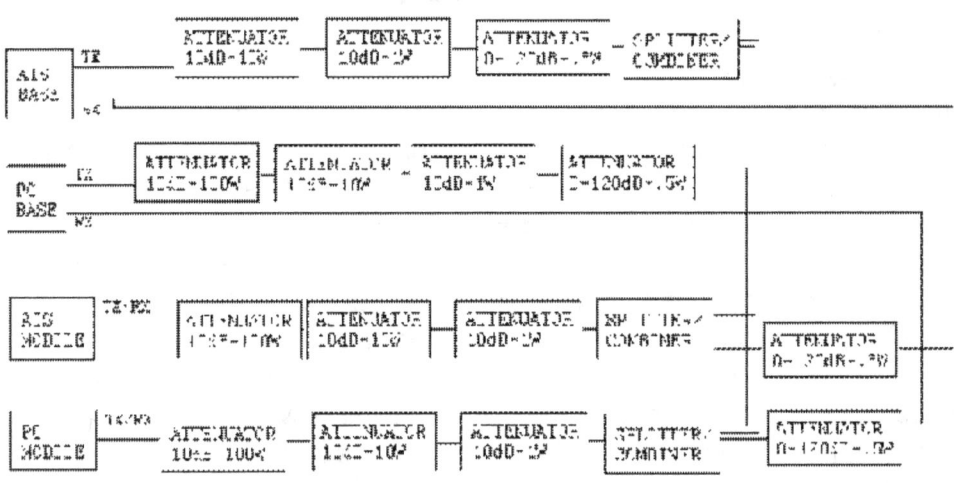

Figure 2-1
Scenario 1 Closed Loop Test Set-up

The following steps were taken to perform the tests.

1. The PC mobile radio and AIS transponder were placed inside a van and connected to the PC and AIS base stations using RF cables coming out of a small access hole in the back of the communications hut. The transponder supplied position reports to the AIS base station, which were then re-transmitted on the channel adjacent to the PC base station.

2. The attenuator on the PC base station was used to vary its RF output power so that the received desired signal, S, at the PC mobile could be changed. The desired signal at the RF input of the radio was set to -60, -95, -101, -107, -110, and -113 dBm. SINAD measurements were taken at each power level. The PC base station was modulated by a 1 kHz tone adjusted in amplitude to produce a 3 kHz deviation.

3. The attenuator on the AIS base station was used to vary its RF output power so that the received interference power, I, at the PC mobile radio could be changed. The VTC operator set the transponder report rate to one second.

4. The desired signal power at the RF input of the radio was set to each value and the power of I was adjusted so that the SINAD was 14 dB. The power of I was recorded into the test log.

5. Step four was repeated for frequency separations of 12.5, 25, 37.5, 50, and 62.5 kHz between

the AIS base station transmitter and the PC mobile receiver.

2.2.2 Scenario 1 Radiated Test Procedures

The following procedures were used in the Scenario 1 radiated tests. A diagram of the test set-up is shown below in Figure 2-2.

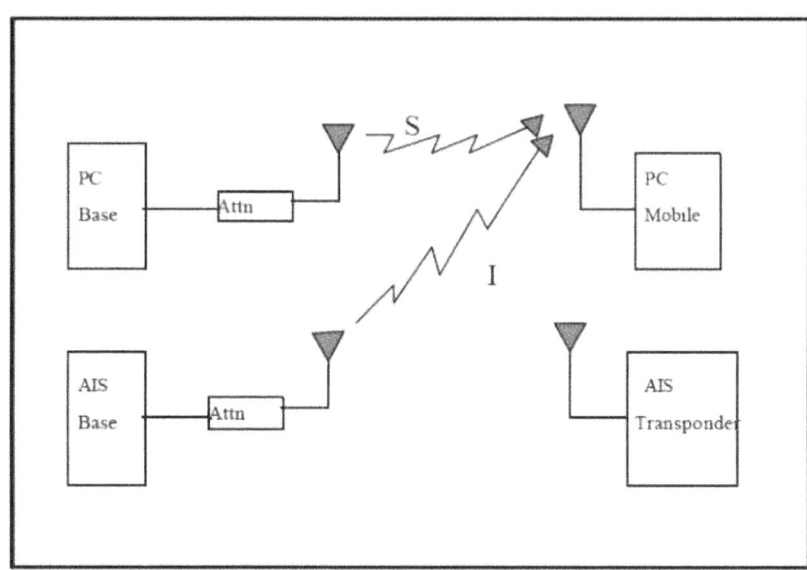

Figure 2-2
Scenario 1 Radiated Test Set-up

1. Two PC mobile radios were placed in a van located 1 mile from the PC and the AIS base station. An AIS transponder was placed outside the communication hut. The PC mobile radio were connected to a 6 dBi gain whip antenna mounted on top of the van.

2. The desired signal power of the PC base station at the mobile receiver was adjusted using the attenuators in the base station to set the mobile receiver SINAD to 14 dB without the AIS base station interference being present. The power level of the PC base station was measured at the PC mobile RF input and recorded into the test log.

3. The power of the AIS base station was then adjusted to make the SINAD fall below 14 dB. The power level of the AIS base station was measured at the PC mobile RF input and recorded into the test log. The frequency separation was 12.5 kHz.

4. The aural quality of the mobile radio receiver was judged by modulating the PC base station with phonetically balanced phrases (adjusted in amplitude to match a 1 kHz tone for a 3 kHz deviation) and listening for bursts of interference from the AIS base station.

2.3 Scenario 2
2.3.1 Scenario 2 Closed Loop Test Procedures

An AIS transponder receiver was tested for susceptibility to interference from a Public Correspondence base station transmitter's emissions using the following procedures. A test set-up diagram is shown below in Figure 2-3.

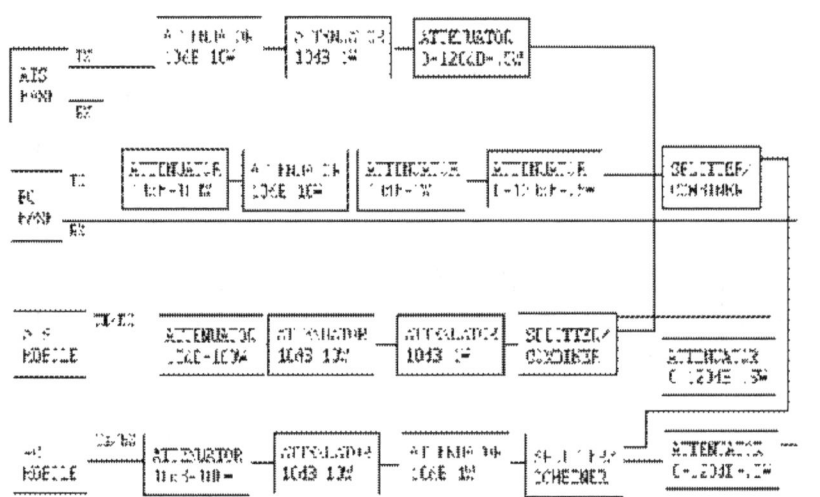

Figure 2-3
Scenario 2 Closed Loop Test Set-up

The following steps were taken to perform the tests.

1. The PC mobile radio and AIS transponder were placed inside a van and connected to the PC and AIS base stations using RF cables.

2. The VTS operator instructed the transponder to report at five second intervals. A second transponder was placed near the tower being tested. This second transponder was used to supply the AIS base station with position reports. The transponder in the van was tested by monitoring the position reports of the second transponder as they were re-broadcast by the AIS base station.

3. The attenuator on the AIS base station was used to vary its RF output power so that the received desired signal, S, at the transponder could be changed. The desired signal at the RF input of the transponder was set to -60, -95, -101, -107, and -110 dBm. At each power level, baseline received report counts were measured and recorded.

4. The attenuator on the PC base station was used to vary its RF output power so that the received interference power, I, at the transponder could be changed. The PC base station transmitter was modulated by voice shaped noise (VSN) adjusted in amplitude to match a 1 kHz tone for a 2.5 kHz deviation.

5. The transponder operator selected the icon of the second transponder on his display and monitored its report rates for a fixed amount of time. The desired signal power at the RF input of the transponder was set to each value and the power of I was adjusted so that the received report count was 90 %. The power of I was recorded into the test log.

6. Step five was repeated for frequency separations of 12.5, and 37.5 kHz between the PC base station transmitter and the AIS transponder receiver. The PC base station was then modulated by a 400 Hz tone adjusted in amplitude to produce a 3 kHz deviation (this was an IEC test requirement) and the tests were repeated.

2.3.2 Scenario 2 Radiated Test Procedures

The following procedures were used in the Scenario 2 radiated tests. A diagram of the test set-up is shown below in Figure 2-4.

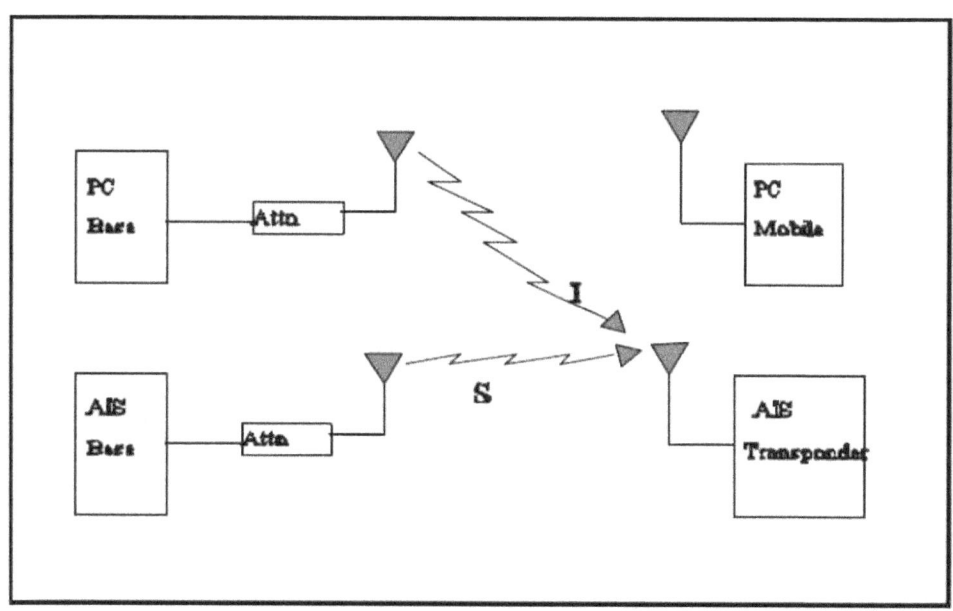

Figure 2-4
Scenario 2 Radiated Test Set-up

1. The van transponder was placed 1 mile from the PC base station and the AIS base station. Another AIS transponder was placed outside the communication hut and logged into the system on

the working channel.

2. The operator of the transponder clicked on the displayed icon of the transponder located at the communications hut and verified that they could receive its position reports which were re-broadcast by the AIS base station.

3. The desired signal power of the AIS base station was adjusted using the attenuators to obtain at least a 90 percent received report count at the transponder without the PC base station interference being present. The power level of the AIS base station was measured at the van and recorded into the test log.

4. The PC base station was then modulated by VSN adjusted in amplitude to match a 1 kHz tone for a 3 kHz deviation. The power of the PC base station was adjusted using attenuators till the received report count at the transponder fell below the baseline measurement.

5. Step four was performed for 12.5 and 37.5 kHz of frequency separation between the PC base station transmitter and the AIS transponder.

2.4 Scenario 3
2.4.1 Scenario 3 Closed loop Test Procedures

A Public Correspondence base station receiver was tested for susceptibility to interference fromn AIS transponder's emissions using the following procedures. A diagram of the test set-up is shown below in Figure 2-5.

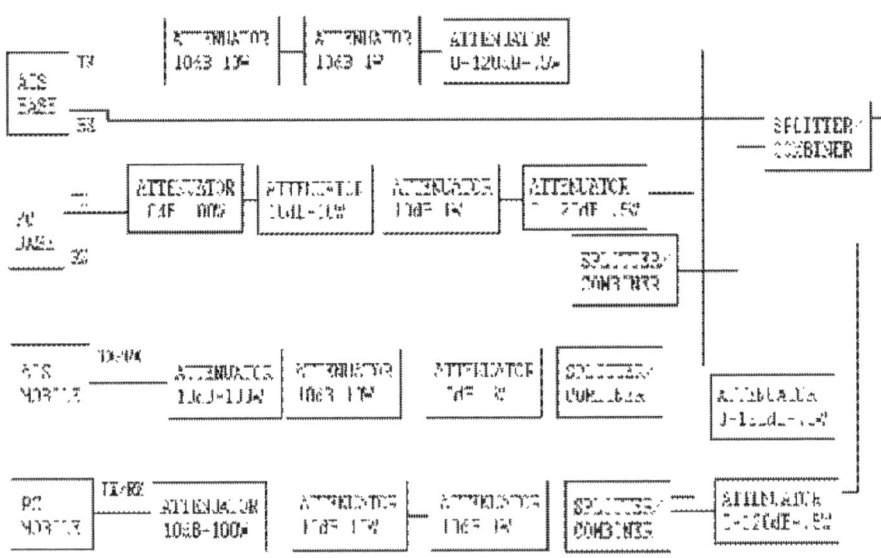

Figure 2-5
Scenario 3 Closed Loop Test Set-up

2-6

The following steps were taken to perform the tests.

1. The PC mobile radio and AIS transponder were placed inside a van and connected to the PC and AIS base stations using RF cables.

2. The VTS operator instructed the transponder to report at 2 second intervals.

3. The attenuator on the PC mobile transmitter was used to vary its RF output power so that the received desired signal power, S, at the PC base station could be changed. The PC mobile radio was modulated by a 1 kHz tone adjusted in amplitude for a 3 kHz deviation. The desired signal power at the RF input of the PC base station receiver was set to -60, -95, -101, -107, -110, and -113 dBm. Baseline SINAD measurements of the PC base station receiver were taken at each desired signal power level before the interference was introduced into the receiver.

4. The attenuator on the transponder was used to vary its RF output power so that the received interference signal, I, at the PC base station receiver could be changed.

5. The desired signal power level was set to each value and the power of I was adjusted so that the SINAD of the PC base station receiver was 14 dB. The power of I was recorded into the test log. The tests were performed with and without the bandpass crystal filter installed on the PC base station receiver. The 3 dB bandwidth of the filter is 25 kHz.

6. Step five was repeated for frequency separations of 12.5, 25, and 37.5 kHz between the PC base station receiver and the AIS transponder.

2.4.2 Scenario 3 Radiated Test Procedures

The procedures that were used in the Scenario 3 radiated tests are described below. A diagram of the test set-up is shown below in Figure 2-6.

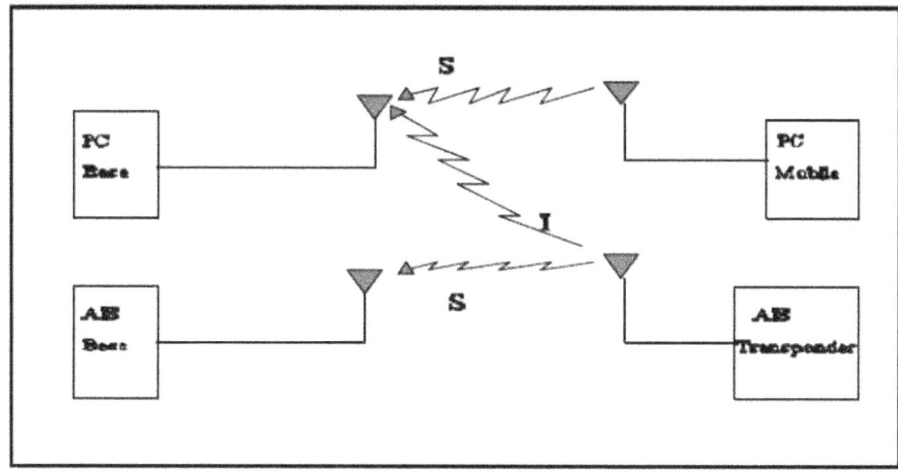

Figure 2-6
Scenario 3 Radiated Test Set-up

1. The PC mobile radio was transmitting with 1 watt output power from a van located 20 miles from the PC base station using an antenna with a 6dBi gain located on the roof. The PC mobile radio was modulated by a 1 kHz tone adjusted in amplitude for a 3 kHz deviation.

2 Personnel at the communications hut set the transponder report rate to two seconds and logged all transponders out of the system except for one located in a car. At the initial starting point, the car was located 1 mile from the PC base station. The transponder was a transportable type and transmitted with 1 watt of output power into a small telescoping whip antenna.

3. Personnel at the communications hut listened to the PC base station receiver for audible interference caused by the transponder's data bursts on adjacent channels.

4. The car containing the AIS transponder was driven away from the PC base station and stopped at distances of 10, 12, 20, 22 and 26 miles and step three was repeated.

This test was performed for frequency separations of 12.5, 25, and 37.5 kHz between the PC base station receiver and the AIS transponder.

2.5 Scenario 4
2.5.1 Scenario 4 Closed Loop Test Procedures
An AIS base station receiver was tested for susceptibility to interference from a Public Correspondence mobile transmitter's emissions using the following procedures. A test set-up diagram is shown below in Figure 2-7.

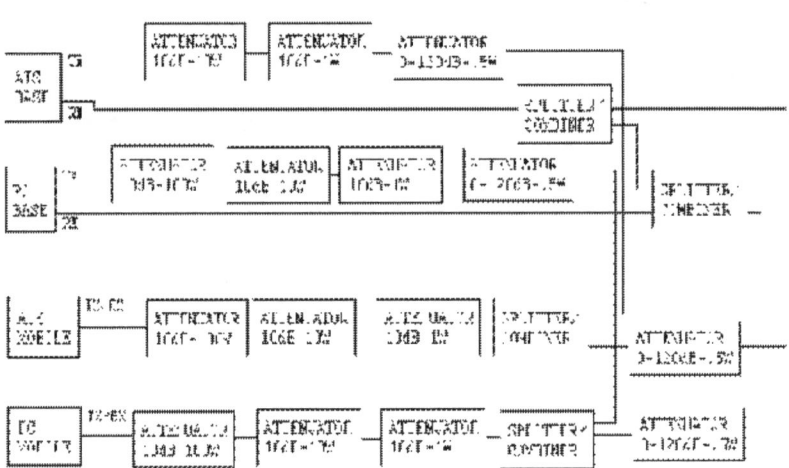

Figure 2-7
Scenario 4 Closed Loop Test Set-up

The following steps were taken to perform the tests.

1. The PC mobile radio and AIS transponder were placed inside a van and connected to the PC and AIS base stations using RF cables.

2. The VTS operator instructed the transponder to report at five second intervals on the working channel.

3. The attenuator on the transponder was used to vary its RF output power so that the received desired signal, S, at the AIS base station receiver could be changed. The desired signal at the RF input to the AIS base station receiver was set to -60, -95, -101, -107, and -110 dBm. Baseline received report counts were taken for the AIS base station receiver at each power level.

4. The attenuator on the PC mobile radio was used to vary its RF output power so that the received interference power, I, at the AIS base station could be changed. The PC mobile radio was modulated by voice shaped noise (VSN) adjusted in amplitude to match a 1 kHz tone for a 2.5 kHz deviation.

5. The VTC operator selected the icon for the transponder with the attenuator on their display and monitored its report rates for a fixed amount of time. The desired signal power at the RF input to the AIS receiver was set to each value and the power of I was adjusted so that the received report count was 90 %. The power of I was recorded into the test log.

6. Step five was repeated for frequency separations of 12.5, 25, and 37.5 kHz between the AIS base station receiver and the PC mobile transmitter. The PC mobile radio was then modulated by a 400 Hz tone adjusted in amplitude to produce a 3 kHz deviation and the tests were repeated.

2.5.2 Scenario 4 Radiated Test Procedures

The procedures that were used in the Scenario 4 radiated tests are described below. A diagram of the test set-up is shown below in Figure 2-8.

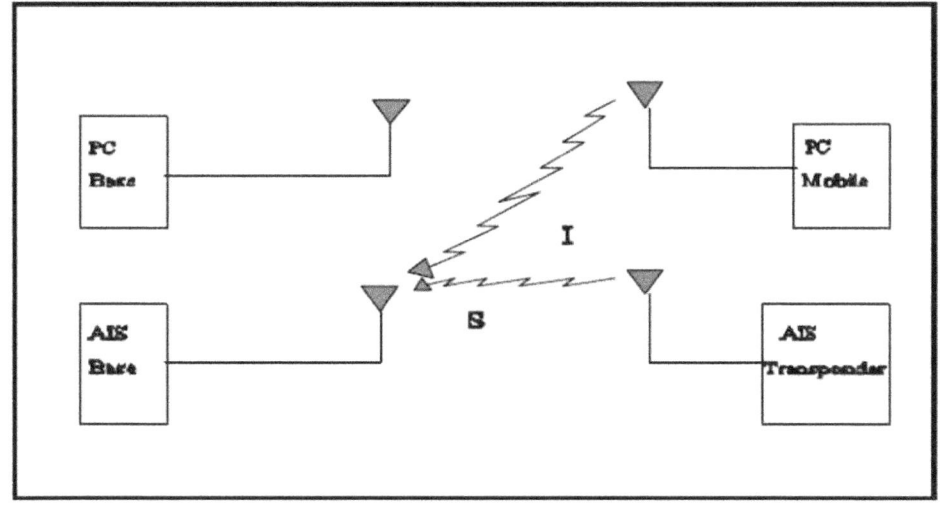

Figure 2-8
Scenario 4 Radiated Test Set-up

1. The AIS transponder was transmitting on the working channel with 1 watt output power at a location 22.4 miles from the AIS base station. Personnel at the communications hut monitored its position reports on the VTC workstation for regular updates via the AIS base station receiver. The report interval was set at five seconds.

2. The van containing a PC mobile transmitter was placed 18, 10, 4.7, 2.5, 1.1, and .5 miles away from the AIS base station. At each location, the PC mobile radio was modulated by VSN adjusted in amplitude to match a 1 kHz tone for a 2.5 kHz deviation and position reports sent from the transponder located 22.4 miles away were observed and counted on the AIS workstation display.

3. Step two was repeated with frequency separations of 12.5, 25, and 37.5 kHz between the AIS base station receiver and the PC mobile transmitter. The power levels at each location were noted and recorded into the test log.

SECTION 3
TEST RESULTS

3. Test Results

The tests are based on the assumption that, for the time being, the AIS will only operate on interstitial channels. The test results made with 25 and 50 kHz of frequency separation between the AIS and PC systems do not reflect an existing interference situation because PC systems are not permitted to operate on interstitial channels at this time. The data points for those frequency separations were taken to add resolution to the graphs.

3.1 Scenario 1

3.1.1 Closed Loop

The susceptibility of PC mobile A's receiver to interference from an AIS base station transmitting on adjacent channels for frequency separations of 12.5 to 62.5 kHz can be determined by reviewing Graph 3-1. The graph shows interference-to-signal (I/S) ratios in dB versus frequency separation for desired signal powers of -101, -107, and -113 dBm for a receiver SINAD of 14 dB. These power levels represent a PC mobile radio operating at the edge of RF coverage for a PC base station tower. For these desired signal power levels and no external interference, the measured SINAD levels were 33, 32, and 28 dB, respectively. They are known as baseline measurements. The Scenario 1 measured data for radio A is contained in Table C-1 of Appendix C.

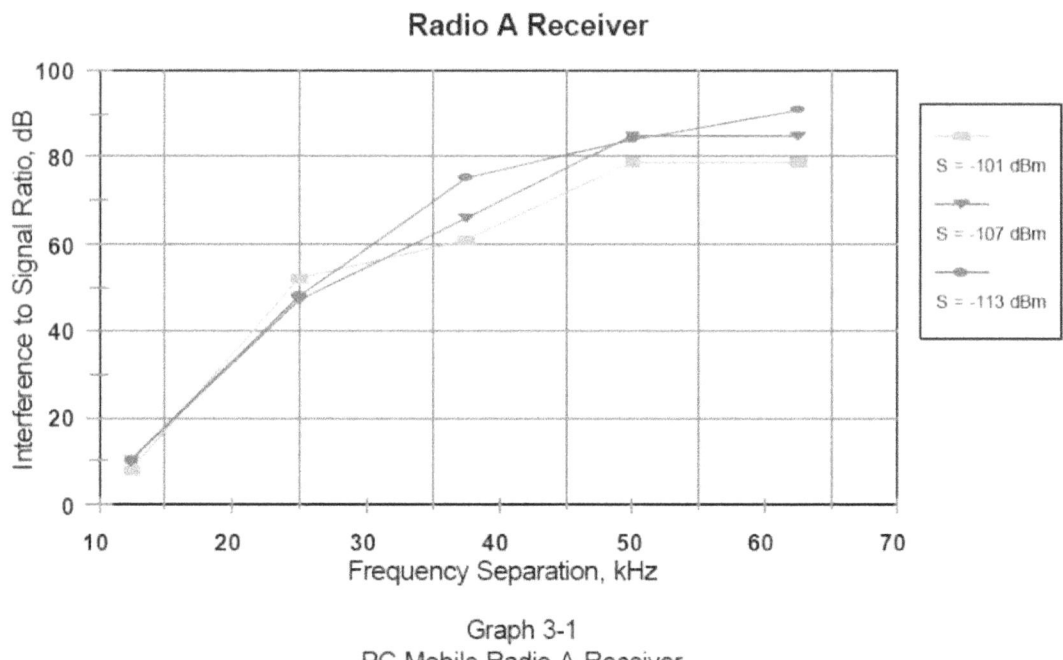

Graph 3-1
PC Mobile Radio A Receiver

Graph 3-1 shows that the I/S ratio increases as the frequency separation between the AIS base station transmitter and the PC mobile receiver increases. A higher I/S ratio means that for a constant desired signal power, more interference power was required to lower the receiver's SINAD to 14 dB as the frequency separation increased. For example, with a desired signal power of -107 dBm the I/S ratio for 12.5 kHz of frequency separation was 10 dB. At 25, 37.5, 50 and 62.5 kHz the I/S ratios are 47, 66, 85 and 85 dB, respectively. Graph 3-1 also shows that beyond 50 kHz of frequency separation the interference rejection capability of receiver A is leveling off.

The susceptibility of PC mobile B's radio receiver to interference from an AIS base station transmitting on adjacent channels for frequency separations of 12.5 to 62.5 kHz can be determined by reviewing Graph 3-2. The graph shows interference-to-signal (I/S) ratios in dB versus frequency separation for desired signal powers of -101, -107, and -113 dBm for a receiver SINAD of 14 dB. The baseline SINAD measurements for these desired signal power levels was 34, 33, and 29 dB, respectively. The Scenario 1 measured data for radio B is contained in Table C-2 of Appendix C.

Radio B Receiver

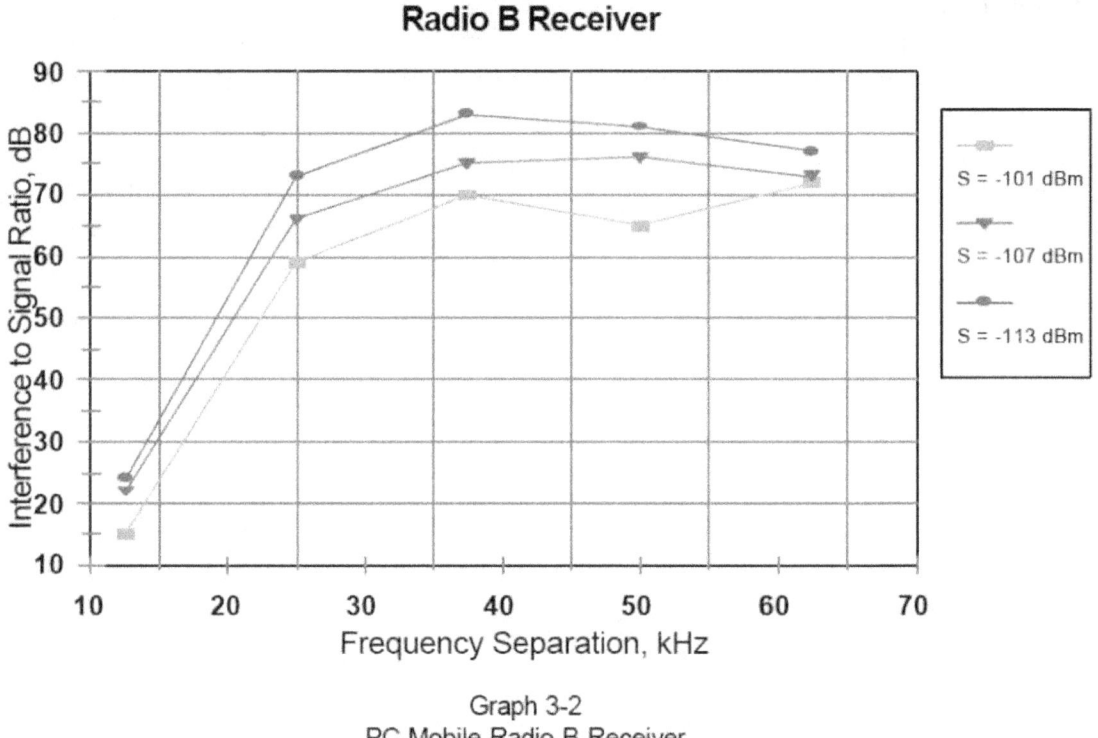

Graph 3-2
PC Mobile Radio B Receiver

Graph 3-2 shows that the I/S ratio increases as the frequency separation between the AIS base station transmitter and the PC mobile receiver increases. A higher I/S ratio means that for a constant desired signal power, more interference power was required to lower the receiver's SINAD to 14 dB as the frequency separation increased. For example, with a desired signal power of -107 dBm the I/S ratio for 12.5 kHz of frequency separation was 22 dB. At 25, 37.5, 50 and 62.5 kHz the I/S ratios are 66, 75, 76 and 73 dB, respectively. Graph 3-2 also shows that beyond 37.5 kHz of frequency separation the interference rejection capability of receiver B is leveling off.

In comparing Graphs 3-1 and 3-2, it can seen that the receiver of radio B is more resistant to adjacent channel AIS interference at frequency separations of 12.5 to 37.5 kHz than receiver A. However, past 37.5 kHz of frequency separation, the receiver of radio A is more resistant to adjacent channel AIS interference by about 10 dB.

3.1.2 Radiated

For the Scenario 1 radiated tests, the van containing the PC mobile radios was driven approximately 1 mile north from the base station and stopped at that location. The PC radios were connected to a roof mounted VHF whip antenna when they were tested. The received desired signal power from the PC base station transmitter and the interference signal power from the AIS base station at the van were controlled by using attenuators at the base station. The frequency separation was 12.5 kHz.

For radio A to achieve a SINAD of 14 dB without the AIS interference present, the desired signal power at the RF input of the radio was -119 dBm. The interference signal power to reduce the SINAD below 14 dB was -98 dBm. This results in a I/S ratio of 21 dB. This is within 6 dB of the I/S ratio for the closed loop tests for a desired signal power of -113 dBm, which was approximately 15 dB. At that level the interference from the AIS base station was barely audible in the radio receiver speaker. To eliminate the audible interference, the AIS power needed to be reduced to -112 dBm at the radio RF input for an I/S of 7 dB. The difference between the audible interference I/S and the I/S for a SINAD of 14 dB is 8 dB.

For radio B to achieve a SINAD of 14 dB without the AIS interference present, the desired signal power at the RF input of the radio was -108 dBm. The interference signal power to reduce the SINAD below 14 dB was -101 dBm. This results in a I/S ratio of 7 dB. At that level the interference from the AIS base station was not audible in the radio receiver speaker. The interference power had to be increased to -92 dBm to make the interference audible in the radio receiver speaker. This results in a I/S ratio of 16 dB. This is within 6 dB of the I/S ratio for the closed loop tests for a desired signal power of -107 dBm which was approximately 22 dB.

3.1.3 Separation Distances

Composite I/S ratios for frequency separations of 12.5, 25, 37.5, and 50 kHz can be developed for a typical VHF radio receiver based on the results of the closed loop tests on radios A and B. The I/S ratios are 15, 60, 75, and 80 dB, respectively. However the tests showed that when the SINAD was 14 dB the interference power had to be lowered by an additional 3 to 14 dB to eliminate the audible interference in the radio receiver Therefore, the I/S ratios are reduced by 5 dB to accommodate the difference between the audible interference and a SINAD measurement of 14 dB. The I/S ratios are then 10, 55, 70, and 75 dB. For frequency separations beyond 50 kHz the level will stay at 75 dB. These I/S ratios are based on a desired signal power of -107 dBm which is equal to 1 • V into a load of 50 ohms.

The maximum AIS base station interference power at the RF input to the PC radio receiver for each frequency separation can then be calculated by adding the I/S ratio to the desired signal power. The results are shown below in Table 3-1.

Table 3-1
Allowable AIS Base Station Interference Power in PC Radio Receiver

Frequency Separation (KHz)	I/S (dB)	PC Base Station Power S, (dBm)	AIS Base Station Power I, (dBm)
12.5	10	-107	-97
25	55	-107	-52
37.5	70	-107	-37
50	75	-107	-32
62.5	75	-107	-32

The distances that correspond to these interference power levels can be determined from previous measurements that were made of the Tower 2 signal strengths. NTIA Report 00-347, "*Lower Mississippi River Ports and Waterways Safety System (PAWSS) RF Coverage Test Results*", contains the measured signal strength for each communications tower referenced in river miles for the AIS and PC base station transmitters. The Tower 2 signal strength graph shown below in Graph 3-3 is referenced in a straight line radial distance. The graph shows that the maximum measured power was -44 dBm. The negative miles are south of the tower and the positive values are north of the tower.

Tower 2
Signal Strength

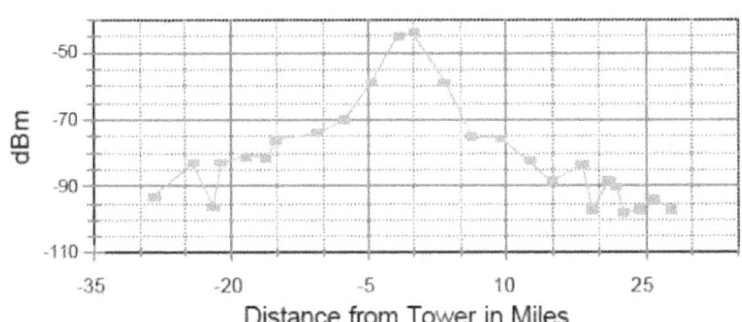

Graph 3-3
Tower 2 Signal Strength

Separation distances for the AIS base station transmitter and the PC mobile radio receiver can be determined from the composite PC mobile radio I/S ratios and Graph 3-3. The PC mobile radio I/S ratios for 12.5, 25, and 37.5 kHz of frequency separation are 10, 55, and 70 dB, respectively. With a desired signal power of -107 dBm the maximum interference powers are then -97, -52, and -37 dBm, respectively. The -97 and -52 dBm power levels correspond to distances of

20 and 2.5 miles. A power level of -37 dBm is under 1 mile.

3.2 Scenario 2

3.2.1 Closed Loop

The I/S ratios for the transponder were calculated using the data collected for Scenario 2 closed loop tests. The results are shown below in Table 3-2.

Table 3-2
Scenario 2 Transponder I/S Ratios for a 90 Percent Received Report Count

Desired Signal Power (dBm)	Interference-to-Signal ratios, dB			
	VSN Modulation		400 Hz Tone Modulation	
	• F = 12.5 KHz	• F = 37.5 KHz	• F = 12.5 KHz	• F = 37.5 KHz
-60	18	35*	13	35*
-95	19	70	16	70*
-101	20	76	17	76*
-107	24	79	19	80
-110	25	81	20	81

*The interference power level was at its maximum value and was not sufficient to bring the received report count below 90 percent.

The results show that the difference between the transponder I/S ratios for the interfering PC base station using VSN modulation versus a 400 Hz tone results were only 2-5 dB.

The results show that the transponder has very good rejection of adjacent channel PC interference for 37.5 kHz of frequency separation. For example, with a desired signal power of -107 dBm the I/S ratio for 12.5 kHz of frequency separation was 24 dB with VSN modulation and 19 dB with tone modulation. For 37.5 kHz of frequency separation the I/S value for VSN was 79 dB and for tone modulation it was 80 dB. This is an improvement of 55 and 61 dB, respectively.

3.2.2 Radiated

In the Scenario 2 radiated tests, the van containing the transponder was driven approximately 1 mile north from the base station and stopped at that location. The transponder was placed on top of the van's roof. The received desired signal power at the van from the AIS base station transmitter and the interference signal power from the PC base station transmitter were controlled by using attenuators at the base station. The test was performed for frequency separations of 12.5 kHz and 37.5 kHz.

The AIS desired signal power was adjusted at the base station to obtain an approximate 90 percent received report count at the transponder without the PC base station interference present. The level was -102 dBm. The PC base station was then modulated with VSN and the received report count was monitored with the interference being on channels 12.5 and 25 kHz adjacent to the working channel.

For 12.5 kHz of frequency separation, the interference power to lower the transponder received report count below 90 percent was -80 dBm. The I/S ratios was approximately 22 dB for the radiated tests with 12.5 khz of frequency separation.

The closed loop tests I/S ratio for 12.5 kHz of separation and a desired signal power of -101 dBm was 20 dB. This is within 2 dB of the radiated test I/S ratio of 22 dB.

3.2.3 Separation Distances

Separation distances for the AIS transponder and the PC base station transmitter can be determined from the transponder I/S ratios and Graph 3-3. The transponder I/S ratios for 12.5, 25, and 37.5 kHz of frequency separation are 22, 50, and 80 dB, respectively. With a desired signal power of -107 dBm the maximum interference powers are then -85, -57, and -27 dBm, respectively. The -85 and -57 dBm power levels correspond to distances of 12 and 3 miles. A power level of -27 dBm is greater than the maximum level that was measured for tower 2 PC base station transmitter and for the other four tower's PC base stations as well.

3.3 Scenario 3

3.3.1 Closed Loop

The susceptibility of the PC base station receiver to interference from an AIS transponder transmitting on adjacent channels for frequency separations of 12.5, 37.5, and 50 kHz can be determined by reviewing Graph 3-4. The graph shows interference-to-signal (I/S) ratios in dB versus frequency separation for desired signal powers of -101, -107, and -113 dBm for a receiver SINAD of 14 dB. The baseline SINAD measurements for these desired signal power levels was a minimum of 30 dB. The measured data for this scenario is contained in Table C-4 of Appendix C.

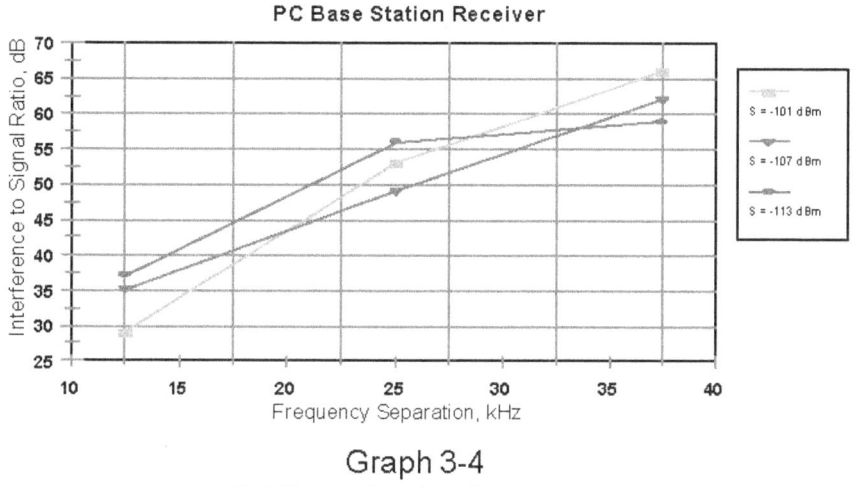

Graph 3-4
PC Base Station Receiver

Graph 3-4 shows that the I/S ratio increases as the frequency separation between an AIS transponder transmitter and the PC base station receiver increases. A higher I/S ratio means that for a constant desired signal power, more interference power was required to lower the receiver's SINAD to 14 dB as the frequency separation increased. For example, with a desired signal power of -107 dBm, the I/S ratio for 12.5 kHz of frequency separation was 35 dB. At 25 and 37.5 kHz of separation, the I/S ratios are 48 and 64 dB, respectively.

3.3.2 Radiated

In the Scenario 3 radiated tests, a PC mobile radio was transmitting from a fixed location 20 miles from the base station. The PC mobile radio was modulated by a 1 kHz tone and was transmitting with 1 watt of power into a 6 dBi gain antenna. Another vehicle containing an AIS transponder was driven away from the base station. Personnel at the communications hut listened for audible interference in the PC base station receiver as the transponder transmitted on channels 12.5, 25, and 37.5 kHz removed from the PC channel.

The results are shown below in Table 3-3. For each frequency separation and distance, the table indicates if the interference was audible in the PC base station receiver speaker.

Table 3-3
Scenario 3 Radiated Tests for Audible Interference in Base Station Receiver

Distance (miles) Transponder to PC base station	• f		
	12.5 kHz	25 kHz	37.5 kHz
10	yes	yes	no
12	yes	yes	no

20	yes	no	no
22	yes	no	no
26	no	no	no

Table 3-3 shows that, with 12.5 kHz of frequency separation, when the transponder was located 26 miles away from the PC base station the audible interference in its receiver was eliminated. At a distance of 22 miles the interference was audible. Therefore, somewhere between 22 and 26 miles is the point where the audible interference would begin to be heard for 12.5 kHz of frequency separation.

For 25 kHz of frequency separation the interference was audible at 12 miles but not at 20 miles. Therefore, somewhere between 12 and 20 miles is the point where the audible interference would begin to be heard for 25 kHz of frequency separation.

For 37.5 kHz of frequency separation, the interference was not audible at a distance of ten miles. Therefore, somewhere closer than 10 miles is the point where the audible interference would begin to be heard for 37.5 kHz of frequency separation. In all cases the PC mobile radio was stationary at a distance of 20 miles from the base station. These results verify the closed loop test data that additional frequency separation offers additional protection for the PC base station receiver.

3.3.3 Separation Distances

The closed loop performance goal was a SINAD of 14 dB while the radiated test performance goal was no audible interference in the PC base station receiver. The radiated tests performance goal was more stringent than the closed loop tests performance goal. In comparing the two, the audible interference was still perceptible when the SINAD was 14 dB. However, the 1 kHz modulation tone could still be heard at all times even during the bursts of interference. The tests revealed that, for an interference power level to lower the SINAD to 14 dB, the same level had to be lowered an additional 6 to 12 dB to eliminate the audible interference.

To further define the distances in Table 3-3 for 12.5 and 25 kHz of frequency separation, a distance corresponding to 12 dB of additional isolation can be added to the last distance at which the interference was heard. This will give an estimate for the separation distances for no audible transponder interference in the PC base station receiver. Graph 3-3 shows that 12 dB of isolation is equivalent to about 3 miles. The geographical separation distance for 12.5 kHz of frequency separation is estimated to be 25 miles and the distance for 25 kHz of frequency separation is estimated to be 15 miles.

The location where the transponder interference could be heard in the PC base station receiver was not found for 37.5 kHz of frequency separation in the radiated tests. However, Graph 3-4 can be used to estimate the geographical separation distance for a frequency separation of 37.5 kHz. Graph 3-4 shows that, for a desired signal power of -101 dBm, the PC base station receiver has 16 dB more protection from transponder interference with a frequency separation of 37.5 kHz than a frequency separation of 25 kHz. Graph 3-3 shows that 16 dB of isolation is equivalent to about 13 miles. Subtracting 13 miles from the 25 kHz separation distance results in 2 miles. Therefore, the geographical separation distance for a PC base station receiver and an AIS transponder for 37.5

kHz of frequency separation is estimated to be about 2 miles. The geographical separation distances would decrease if the desired signal power were increased.

These tests were conducted with and without the bandpass crystal filter inserted in the RF path leading to the PC base station receiver input. The test results showed that, for this interference scenario, the crystal filter had a negligible effect on the receivers ability to reject adjacent channel AIS interference. The crystal filter is designed for 25 kHz channel operations and it only shows 5 dB of insertion loss at 12.5 kHz off-tuned from its center frequency. At 25 kHz off-tuned from its center frequency the filter had about 29 dB of rejection in addition to its 5 dB of insertion loss.

These separation distances are based on the technical characteristics of the PC base station receiver that was tested and its antenna configuration. Other PC base stations may require slightly different separation distances to account for differences in antenna configuration/installation and RF circuitry/amplifiers. An important factor is the IF bandwidth of the receiver. If the receiver were optimized for 12.5 kHz channelized operation its IF bandwidth would be about 9 to 10 kHz. The current IF bandwidth for operation with 25 kHz channelization is set at about 15 kHz. A narrower IF bandwidth would make the receiver less susceptible to adjacent channel interference and thus lower the separation distances.

3.4 Scenario 4
3.4.1 Closed Loop

The susceptibility of the AIS base station receiver to interference from a PC mobile radio transmitting on adjacent channels for frequency separations of 12.5, 25, and 37.5 kHz can be determined by reviewing Graph 3-5. The graph shows interference-to-signal (I/S) ratios in dB versus frequency separation for desired signal powers of -101, -107, and -110 dBm for a received report count of 90 percent. The baseline received report count for these desired signal power levels was near 100 percent. The measured data for this scenario is contained in Table C-5 of Appendix C.

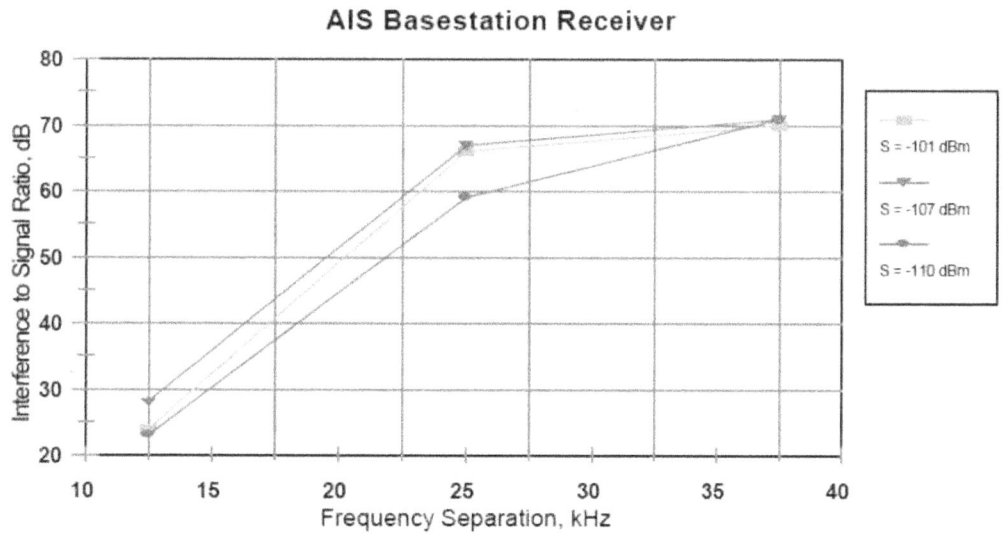

Graph 3-5
AIS Base Station Receiver

3-9

Graph 3-5 shows that the I/S ratio increases as the frequency separation between a PC mobile transmitter and the AIS base station receiver increases. A higher I/S ratio means that for a constant desired signal power, more interference power was required to lower the received report count to 90 percent as the frequency separation increased. For example, with a desired signal power of -107 dBm the I/S ratio for 12.5 kHz of frequency separation was 28 dB. At 25 and 37.5 kHz of separation the I/S ratios are 67 and 71 dB respectively.

Graph 3-5 also shows that beyond 25 kHz of frequency separation, the I/S ratios do not significantly increase for the AIS base station receiver. For desired signal powers of -101 and -107 dBm the difference between the I/S ratios for 25 and 37.5 kHz of frequency separation is only 5 dB. Graph 3-5 also shows that beyond 25 kHz of frequency separation the interference rejection capability of the AIS base station receiver is leveling off.

3.4.2 Radiated

The results of the radiated tests for Scenario 4 are summarized in Table 3-4. For this test a transportable AIS transponder was located 22.4 miles from the AIS base station and transmitting on low power (1 watt) on the working channel with a small telescoping antenna. At a distance of 22.4 miles, the power of the AIS transponder was calculated to be -107 dBm at the RF input to the AIS base station receiver The transponder remained fixed at that location for the duration of this test and the van containing a PC mobile radio transmitter was stopped at different distances from the base station. At each location, the PC mobile radio was modulated with VSN on a channel adjacent to the transponder working channel. The effects of the PC mobile radio interference on the AIS base stations received report count for the stationary transponder were observed on the VTC workstation display. Without the PC mobile interference present, the received report count for the stationary transponder was at least 95 percent at all times. Table 3-4 shows the percent of the received report count for frequency separations of 12.5, 25, and 37.5 kHz.

Table 3-4
Scenario 4 Radiated Test Results

Distance (miles) PC mobile to AIS base station	•f		
	12.5 kHz	25 kHz	37.5 kHz
.5	Below 10 %	• 85%	• 95%
1.1	Below 10 %	• 85%	• 95%
2.5	Below 10 %	• 87%	• 95%
4.7	Below 10 %	• 87%	• 95%
10	• 95%	• 95%	• 95%
18	• 95%	• 95%	• 95%

For this test, the AIS base station was connected to tower transmit and receive antennas in addition to the RF amplifiers inside the RF cabinet. The PC mobile radio was transmitting with 1 watt of output power into a 6dBi gain whip antenna mounted on the roof of the van.

The data in Table 3-4 shows that the PC mobile transmitter did not have an effect on the received report count for 12.5 kHz of separation until it was 4.7 miles from the base station. At that location with 12.5 kHz of frequency separation the received report count was below 10 percent. With 10 miles of geographical separation and 12.5 kHz of frequency separation the received report count was above 95 percent. Therefore, somewhere between 4.7 and 10 miles is the point where the received report would be reduced to 90 percent for 12.5 kHz of frequency separation.

For 25 kHz of frequency separation and 4.7 miles of geographical separation, the PC mobile radio had some effect on the received report count, but it never went below 85 percent as the geographical separation decreased. For 37.5 kHz of frequency separation, the PC mobile did not have any effect on the AIS base station received report count at the closest distance of .5 miles.

The results of the radiated tests confirm the results of the closed loop tests which show that, for frequency separations of 37.5, the AIS base station receiver is very resistant to adjacent channel PC interference.

3.4.3 Separation Distances

The results of the radiated and closed loop tests for Scenario 4 show that the AIS base station receiver will not be affected by a PC mobile radio operating 12.5 kHz away from the working channel as long as the PC mobile radio was at a minimum 10 miles from the AIS base station. For a frequency separation of 25 kHz, the PC mobile radio would have a small effect on the AIS base station receiver at distances up to .5 miles. For a frequency separation of 37.5 kHz, the PC mobile radio would not have any effect on the AIS base station receiver at distances up to .5 miles.

These distances are based on a PC mobile radio transmitter power of 1 watt. The PC mobile radios are allowed to transmit with a maximum power of 25 watts. The geographical separation distance that was calculated for 12.5 kHz of frequency separation must be adjusted by 14 dB to take into account the difference between high power and low power PC mobile radio operations. Graph 3-3 shows that 14 dB of isolation is equivalent to about 5 miles. Therefore, for a frequency separation of 12.5 kHz and the PC mobile radio transmitting on high power, the separation distance between the AIS base station receiver and the PC mobile radio is 15 miles. This assumes a transponder received desired signal power of -107 dBm.

Graph 3-5 shows that the I/S ratio for 25 kHz of frequency separation is about 67 dB, which is 39 dB greater than the 12.5 kHz frequency separation I/S ratio of 28 dB. Graph 3-3 shows that 39 dB of isolation is equivalent to 11 miles. Therefore, for a frequency separation of 25 kHz and the PC mobile radio transmitting on high power, the separation distance between the AIS base station receiver and the PC mobile radio is 4 miles. This assumes a transponder received desired signal power of -107 dBm.

Graph 3-5 shows that the I/S ratio for 37.5 kHz of frequency separation is about 71 dB, which is 43 dB greater than the 12.5 kHz frequency separation I/S ratio of 28 dB. Graph 3-3 shows that 43 dB of isolation is equivalent to 14 miles. Therefore, for a frequency separation of 37.5 kHz and the PC mobile radio transmitting on high power, the separation distance between the AIS base station receiver and the PC mobile radio is 1 mile. This assumes a transponder received desired signal power of -107 dBm.

These separation distances are based on the technical characteristics of the AIS base station receiver that was tested and its antenna configuration. Other AIS base stations may require slightly different separation distances to account for differences in antenna configuration/installation and RF circuitry/amplifiers.

SECTION 4
SEPARATION DISTANCES

4. Separation Distances

Separation distances were calculated for each of the four test scenarios described in Section 1 of the report. The tests are based on the assumption that the AIS (based on ITU Recommendation M. 825-3) will operate on 12.5 kHz interstitial channels and the Public Correspondence will operate on 25 kHz channels The results are shown below. However, it should be noted that tests made with 25 kHz of frequency separation between the AIS and PC systems do not reflect an existing interference situation because PC systems are not permitted to operate on interstitial channels at this time. These tests were conducted for data trends purposes and are not directly applicable as such operations would require a regulatory change in existing wideband Public Correspondence operations.

The test results show that, considering all four receivers, the PC base station receiver is the most susceptible to adjacent channel interference and requires the largest geographic separation distance for protection. For 12.5 kHz of frequency separation, the PC base station receiver requires 25 miles of geographic separation from an AIS transponder. Since PC and AIS systems would employ base stations to serve mobile units on vessels in the same geographic area, operations of these systems in the same geographic area may not be practical with 12.5 kHz of frequency separation. A detailed summary of the separation distances is given in the following paragraphs.

4.1 Base Station Receiver vs Mobile Interferer

The separation distances for the PC base station receiver and an AIS transponder (Scenario 3) are shown below in Table 4-1 for frequency separations of 12.5, 25, and 37.5 kHz. To protect the PC base station receiver from transponder interference, the AIS transponder should not operate within the area that is defined by a circle which has a radius equal to the separation distance that is shown in Table 4-1 for each frequency separation. These distances are based on a received desired signal power of -98 dBm at the RF input of the PC base station receiver. This power level corresponds to a fixed mount radio transmitting from distance of twenty miles from a PC base station, which was considered a nominal distance for PC communications. The separation distances would be reduced if the desired signal power is increased.

Table 4-1
Scenario 3
PC Base Station Receiver and AIS Transponder Separation Distances

Frequency Separation (kHz)	Distance (miles)
12.5	25
25	15
37.5	2

The separation distances for the AIS base station receiver and an PC mobile radio (Scenario 4) are shown below in Table 4-2 for frequency separations of 12.5, 25, and 37.5 kHz. To protect the AIS base station receiver from PC mobile radio interference, the PC mobile radio should not operate within the area that is defined by a circle which has a radius equal to the separation distance that is shown in Table 4-2 for each frequency separation. These distances are based on a received desired signal power of -107 dBm at the RF input of the AIS base station receiver. The distances would be reduced if the desired signal power is increased.

Table 4-2
Scenario 4
AIS Base Station Receiver and PC Radio Separation Distances

Frequency Separation (kHz)	Distance (miles)
12.5	15
25	4
37.5	1

4.2 Mobile Receiver vs. Base Station Interferer

The separation distances for the PC mobile receiver and an AIS base station transmitter (Scenario 1) are shown below in Table 4-3 for frequency separations of 12.5, 25, and 37.5 kHz. To protect the PC mobile radio from AIS base station interference, the PC mobile radio should not come any closer to the AIS base station than the distances that are shown in Table 4-3 for each frequency separation. These distances are based on a received desired signal power of -107 dBm at the RF input of the PC mobile radio receiver. The distances would be reduced if the desired signal power is increased.

Table 4-3
Scenario 1
PC Mobile Radio Receiver and AIS Base Station Separation Distances

Frequency Separation (kHz)	Distance (miles)
12.5	20
25	2.5
37.5	1

The separation distances for the AIS transponder and a PC base station transmitter (Scenario 2) are shown below in Table 4-4 for frequency separations of 12.5, 25, and 37.5 kHz. To protect the transponder receiver from PC base station interference, the transponder should not come any closer to the PC base station than the distances that are shown in Table 4-4 for each frequency separation. These distances are based on a received desired signal power of -107 dBm at the RF input of the AIS transponder. The distances would be reduced if the desired signal power is increased.

Table 4-4
Scenario 2
AIS Transponder Receiver and PC Base Station Separation Distances

Frequency Separation (kHz)	Distance (miles)
12.5	12
25	3
37.5	.5

4.2 Considerations for ITU-RM1371 Standard

The new AIS standard, recommendation ITU-R M.1371, has a higher data rate and a different modulation scheme than the ITU-R M.825-3 standard, which is the current AIS protocol being used by the Coast Guard. The new standard's higher data rate and different modulation scheme will require that the transponder and AIS base stations need slightly larger separation distances from the PC systems for the same level of protection.

SECTION 5
CONCLUSIONS and RECOMMENDATIONS

5.1 General Conclusions

Tests scenarios were designed to determine compatibility between AIS operating on interstitial channels (12.5kHz) and PC operating on 25 kHz channels operating within the VHF Maritime Mobile Band using duplex channels. AIS operations conformed to ITU Recommendation ITU-R M. 825-3.

A minimum 12.5 kHz of frequency separation was used between the interfering transmitter and victim receiver in all test scenarios. In addition, dependent on the test scenario being used, 25 kHz, 37.5 kHz, 50 kHz and 62.5 kHz frequency separations were also tested. Tests using 25 kHz and 50 kHz frequency separations were conducted for data trends purposes and are not directly applicable as they would require regulatory changes applicable to PC operations.

The four scenarios tested were: 1) An AIS base station causing interference to a PC mobile radio receiver, 2) a PC base station causing interference in a AIS transponder receiver, 3) an AIS transponder causing interference in a PC base station receiver, and 4) a PC mobile radio causing interference to an AIS base station receiver. A fifth test scenario required to determine compatibility between ship-to-ship AIS operations and PC operations was not tested due to the unavailability of suitable equipment. This fifth scenario represents a co-site concern with both AIS and PC systems operating on the same ship. The inability to test this fifth scenario does not alter the conclusions or recommendations given in this report.

Analyses of results obtained for the four test scenarios offer general guidelines for determining compatibility between AIS operations on interstitial 12.5 kHz channels and Public Correspondence operations on wideband 25 kHz channels. Test results are specific to the New Orleans area and are generally applicable to other areas as well.

5.2 Specific Conclusions

Based on the previously established criteria and analyses of the test results, the following is concluded:

1) Since AIS and PC systems both offer service to mariners on ships and would employ base stations with transmission towers located in the same geographic environment, operating these systems in the same area may not be practical with 12.5 kHz of frequency separation (i.e., geographical separation distance on the order of 20 miles is required).

2) The PC and AIS systems should be able to operate within the same geographic environment provided that a minimum of 37.5 kHz of frequency separation between the two systems is maintained.

5.3 Recommendations

NTIA recommends that the Coast Guard consider:

1. Developing an AIS frequency coordination plan for the lower Mississippi River for the PC and AIS systems that will ensure mutually compatible and satisfactory operations.

2. Performing additional EMC tests between ship-to-ship AIS and PC operations.

3. Performing EMC tests between PC systems and ITU-R. M 1371 compliant AIS equipment when such equipment becomes available.

4. Pursuing necessary regulatory changes to improve AIS and PC operations in the same geographical area (e.g., including a 12.5 kHz channelization for both AIS and PC operations and developing appropriate receiver standards).